GRAPHIC CAREERS
FIGHTER PILOTS

by David West

illustrated by James Field

rosen publishing's
rosen central

New York

Published in 2008 by The Rosen Publishing Group, Inc.
29 East 21st Street, New York, NY 10010

First edition, 2008

Designed and produced by
David West Books

Editor: Gail Bushnell

Photo credits:
4b, U.S. Air Force; 5b, U.S. Air Force; 6b, Adrian Pingstone; 7tr, U.S. Navy photo by Journalist 1st Class David Rush; 7mr, U.S Navy photo by Photographer's Mate 3rd Class Staci M. Bitzer; 7bl, U.S. Air Force photo/Staff Sgt. Aaron D. Allmon II; 45, U.S Air Force photo by Ken Hackman

Library of Congress Cataloging-in-Publication Data

West, David.
  Fighter pilots / by David West ; illustrated by James Field.
     p. cm. -- (Graphic careers)
  Includes index.
  ISBN 978-1-4042-1455-2 (library binding) -- ISBN 978-1-4042-1456-9
(pbk.) -- ISBN 978-1-4042-1457-6 (6 pack)
  1. Fighter pilots--Juvenile literature. 2. Fighter plane
combat--United States--Juvenile literature. 3.
Airplanes--Piloting--Juvenile literature. I. Title.
  UG703.W47 2000
  358.4'302373--dc22

                              2007041458

Manufactured in China

CONTENTS

RISE OF THE FIGHTER

A fighter plane is designed to attack other aircraft, as opposed to a bomber, which is designed to attack ground targets by dropping bombs.

Manfred Albrecht Freiherr von Richthofen (May 2, 1892–April 21, 1918) was a German fighter pilot known as the Red Baron. He was the most successful flying ace of World War I, and was credited with 80 confirmed air combat victories.

WORLD WAR I

The term "fighter" was not used until after World War I, even though the war gave birth to single-seater fighters and the fighter ace (a pilot who shoots down five or more planes). Fighters were developed in response to aircraft being used for bombing and reconnaissance. The invention of the synchronizer gear allowed machine guns to fire through the propeller space without hitting the propeller. Aerial warfare became more important as each side wrestled for control over the air.

Developed in April 1915, the Fokker Eindecker was the first German aircraft built specifically to be a fighter plane, and the first aircraft fitted with synchronizer gear.

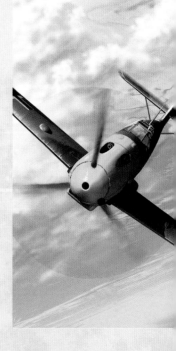

Edward Vernon Rickenbacker was the American "Ace of Aces" in World War I. Born in Columbus, Ohio, in 1890, he gained fame as a race car driver before joining the service.

WORLD WAR II

By World War II, fighters were mainly all-metal monoplanes with heavy machine guns or cannons mounted in the wings. Epic battles in the skies over Europe saw fighter against bomber and fighter against fighter. The war at sea gave rise to the carrier-borne fighters. The war in the Pacific witnessed some of the most fearsome air-to-air combat between the Japanese Zero and the American Hellcat. In some cases the role of the fighter was merged with the role of the bomber, resulting in the first of the dual-role fighters, that could attack ground targets as well as take on enemy fighters.

Grumman F6F Hellcat

Mitsubishi A6M Zero

This artist's rendering shows British Spitfire and Hurricane fighters engaged in dogfights with Messerschmitt Me-109s and Heinkel bombers in the skies over London, England, during the Battle of Britain, in 1940.

A Republic P-47D (right) in flight, firing rockets. On the way back from escorting bombers on raids, pilots shot up targets on the ground, and also used belly shackles to carry bombs on short-range missions. This led to the realization that the P-47 could perform a dual role on escort missions as a fighter-bomber.

JET FIGHTERS

Glocester Meteor

The first jet fighters appeared toward the end of World War II (1939–1945). Both Great Britain and Germany produced jet fighters that entered service in 1944.

Messerschmitt Me-262

FIRST GENERATION JET FIGHTERS (1944–1953)

Britain's Glocester Meteor was powered by two jet engines mounted on the wings, as was the German Messerschmitt Me-262, which had swept wings. The Me-262 was very successful and had a great influence on the design of jet fighters. The end of this period witnessed the first jet-to-jet fighter combats during the Korean War (1950–1953), when American F-86 Sabres tangled with Russian-built MiG-15s.

MiG-15

American F-86 Sabres scored a victory ratio of three-to-one over the North Korean MiG-15s.

SECOND GENERATION JET FIGHTERS (1953–1960)

Many new technologies improved the fighting capabilities of the jet fighter during this period. Air-to-air missiles moved combat beyond visual range, which required the installation of radar to acquire targets. Swept, delta, and variable wing designs were tried and tested. Agility was replaced with speed and strength for a large missile payload, and the world saw new types of jet fighters emerge. These were the fighter-bombers (such as the F-105 and the Sukhoi Su-7), and the interceptor (such as the English Electric Lightning).

RAF pilots flying the English Electric Lightning described it as being "saddled to a skyrocket."

THIRD GENERATION JET FIGHTERS (1960–1970)

F-4 Phantom

This period saw jet fighters in close-quarter dogfights during the Vietnam War (1959–1975), where American F-4 Phantoms battled against North Vietnamese MiG-17s, -19s, and -21s. The result was a return to more agile fighters with guns. Jet fighters also became multi-role aircraft instead of having a special role such as heavy fighter, strike fighter, or interceptor. One of the most unusual fighters, the British Harrier Jump Jet, is truly multi-role in its ability to vertically take off and land (VTOL). This has made it popular with the Navy and Marines. It proved itself as a fighter and bomber during the Falklands War (1982).

F-86 Sabre

The U.S. Navy's Harrier AV-8B

FOURTH GENERATION JET FIGHTERS (1970–2000)

Computer technology and modern avionics made producing a jet fighter extremely expensive. With rising costs, multi-role fighters became more popular. It was during this period that the F-117A Nighthawk introduced stealth technology to the air. It flew 1,300 sorties during the First Gulf War (1990–1991). Other multi-role fighters, such as the European Panavia Tornado, F-15 Eagle, and the F/A-18 Hornet, also saw action during the First Gulf War.

F-117A

FIFTH GENERATION JET FIGHTERS (2000–TODAY)

Today's modern fighter relies on cutting-edge technologies. Composite materials, stealth technology, advanced radar, and computer-aided avionics are used to reduce a pilot's workload while improving his or her awareness. The X-35C Joint Strike Fighter is the latest of these fifth generation jet fighters being tested.

X-35C Joint Strike Fighter

Lt. Edwin C. Parsons
WWI FIGHTER PILOT
The Escadrille Lafayette[*]

IT IS SUMMER 1916 IN FRANCE. WORLD WAR I RAGES ALONG THE TRENCHES OF THE WESTERN FRONT. THE UNITED STATES IS NOT YET IN THE WAR, BUT MANY AMERICANS ARE JOINING THE FRENCH ARMED FORCES TO FIGHT THE GERMANS. EDWIN "TED" PARSONS HAS SIGNED UP FOR FLIGHT TRAINING...

*THE NAME GIVEN TO THE ESCADRILLE AMÉRICAINE (AMERICAN AIR SQUADRON), WHICH WAS CHANGED ON NOVEMBER 6, 1916. IT MEANS THE "LAFAYETTE AIR SQUADRON."

ALTHOUGH HE HAS FLOWN BEFORE, PARSONS LEARNS THE BASICS OF FLYING A PLANE ALONG WITH THE OTHER TRAINEES.

GOOD MORNING, GENTLEMEN. TODAY YOU WILL BE LEARNING THE CONTROLS AND HOW THEY AFFECT THE CONTROL SURFACES OF THE AIRCRAFT.

PUSHING THE PEDAL WITH YOUR LEFT FOOT MAKES THE RUDDER TURN RIGHT.

HEY, TED, I THINK HE MEANS YOUR OTHER LEFT FOOT...

HA HA HA HAHAHA

GENTLEMEN!

MOST PILOTS DO NOT SURVIVE BEYOND THREE WEEKS. I SUGGEST YOU PAY ATTENTION.

EVENTUALLY, THE TRAINEES ARE SET FREE IN A BLERIOT PENGUIN. THE WINGS OF THE PLANE ARE CLIPPED SO THAT IT CANNOT TAKE OFF.

WHOAH!

AFTER HIS FIRST FLIGHT A TRAINEE WAS AWARDED HIS MILITARY PILOT'S CERTIFICATE, OR BREVET, AND PROMOTED TO LANCE CORPORAL.

EVERYONE SAY, "CHEESE."

THE NEW PILOTS NOW HAVE TO LEARN AEROBATICS FOR DEFENSE AND ATTACK.

YOU MUST GAIN AS MANY HOURS OF FLYING AS POSSIBLE WHILE LEARNING THESE AEROBATICS. HERE IS THE MOST SIMPLE OF TURNS THAT YOU WILL LEARN...

...SO, THIS AFTERNOON YOU WILL BE FLYING IN PAIRS AND PERFORMING THE "CHANGE OF DIRECTION."

THERE IS LITTLE TIME FOR THE NOVICES TO LEARN AEROBATICS...

...AND GUNNERY PRACTICE...

RATATATATATAT

...BEFORE THEY ARE CALLED TO THE ESCADRILLES.

LANCE CORPORAL PARSONS, YOU WILL BE GOING TO THE ESCADRILLE LAFAYETTE.

YES, SIR. THANK YOU, SIR.

ON JANUARY 24, 1917, EDWIN PARSONS ARRIVES AT CACHY, NEAR AMIENS, WHERE THE ESCADRILLE LAFAYETTE IS BASED. THE ESCADRILLE'S OFFICIAL NAME IS N124. THE N STANDS FOR NIEUPORT—THE TYPE OF FIGHTER THE SQUADRON FLIES.

WOW. I CAN'T WAIT TO GET INTO ONE OF THOSE BABIES.

THE ESCADRILLE LAFAYETTE IS MOSTLY MADE UP OF AMERICAN VOLUNTEERS. IT HAS A REPUTATION FOR HAVING A PARTY ATMOSPHERE, ALTHOUGH THE FLYING IS TAKEN VERY SERIOUSLY.

YIKES! WHOAH THERE!

GRRRRR

DON'T WORRY. HE'S FRIENDLY ENOUGH. HIS NAME IS WHISKEY, AND HE'S OUR MASCOT.

ON JANUARY 27, THE SQUADRON MOVES TO ST. JUSTE, NEAR RAVENEL.

THE WEATHER IS SO COLD THAT PARSONS SLEEPS WITH WHISKEY FOR WARMTH.

BOY, HAVE YOU GOT BAD BREATH.

YAAAWNRRR

A FEW DAYS AFTER THEIR ARRIVAL AT ST. JUSTE, PARSONS IS TAKEN OUT ON PATROL BY LIEUTENANT WILLIAM THAW.

STICK CLOSE, AND NO HEROICS.

SERGEANT RAOUL LUFBERY HAS JUST RETURNED FROM LEAVE.

WHO'S THAT TAKING OFF, FRED?

THAT'S THAW'S PATROL.

THE PATROL IS BETWEEN THE SOMME AND OISE RIVERS WHEN PARSONS SPOTS FIVE ENEMY PLANES BELOW. HE WAGGLES HIS WINGS AND POINTS OUT THE PLANES TO THAW. THAW SHAKES HIS HEAD...

15

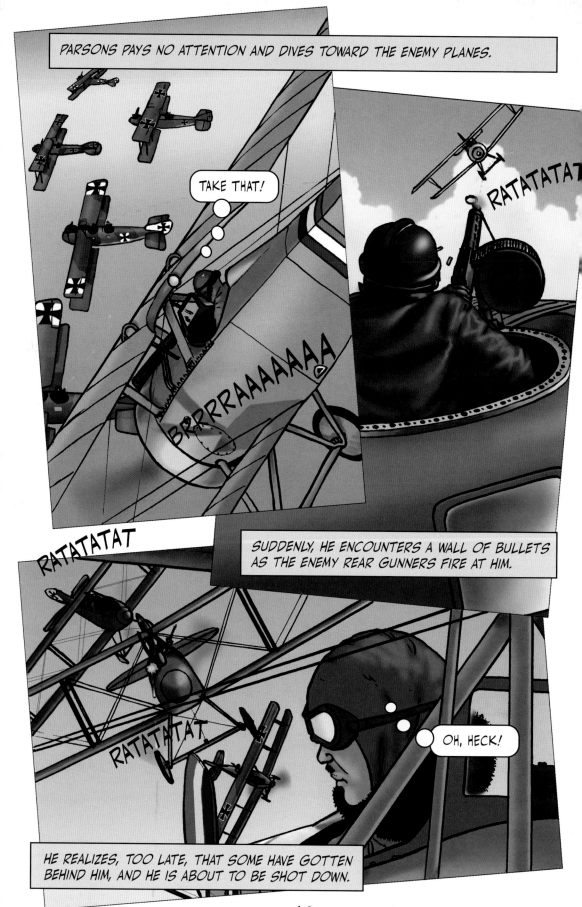

16

HE PULLS BACK INTO A CLIMB, EXPECTING TO BE SHOT TO PIECES AT ANY MOMENT...

ZZZIP

ZZZIP

I'M DONE FOR!

...BUT WHEN HE LOOKS BACK OVER HIS SHOULDER HE SEES...

RATATATAT

THAT LOOKS LIKE LUFBERY. WHERE DID HE COME FROM? HE CERTAINLY SAVED MY SKIN.

...TWO OF THE ENEMY GOING DOWN OUT OF CONTROL, ONE IN FLAMES, AND THE REST SCATTERING BEFORE THE NIEUPORT OF RAOUL LUFBERY. LUFBERY HAD TAKEN OFF SOON AFTER THE PATROL HAD LEFT AND ARRIVED IN TIME TO SAVE PARSONS' LIFE. PARSONS WENT ON TO EVENTUALLY SHOOT DOWN SEVEN ENEMY PLANES, MAKING HIM ONE OF THE FEW ACES THAT FLEW WITH THE ESCADRILLE LAFAYETTE.
THE END

PILOT OFFICER GEOFFREY WELLUM
RAF SPITFIRE PILOT
THE BATTLE OF BRITAIN

IT IS JULY 1940. THE ALLIED FORCES HAVE RETREATED FROM MAINLAND EUROPE. BRITAIN STANDS ALONE. GERMANY IS PREPARING TO INVADE THE ISLAND BUT NEEDS TO CLEAR THE SKIES OF ENEMY AIRCRAFT. FEWER THAN 700 BRITISH FIGHTER PLANES AWAIT THE GERMAN AIR RAIDS. GEOFFREY WELLUM AND THE REST OF NO. 92 SQUADRON WAIT BY THE DISPERSAL HUT CLOSE TO THEIR SPITFIRES AT BIGGIN HILL. THE PHONE RINGS...

OK, OLD BOY. RIGHT AWAY. CHEERS.

SQUADRON SCRAMBLE!*

*THE ORDER TO TAKE OFF IMMEDIATELY.

START UP!

COME ON, GET A MOVE ON!

THE SPITFIRE HAS ALREADY BEEN STARTED BY THE GROUND CREW BY THE TIME WELLUM REACHES HIS PLANE. THEY HELP STRAP HIM INTO HIS COCKPIT.

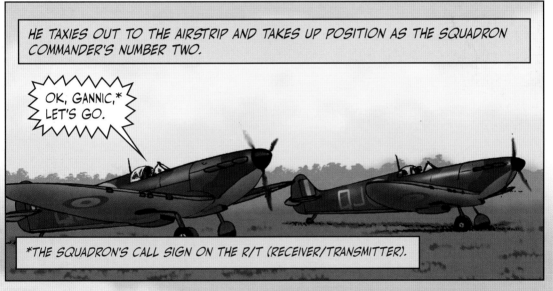

HE TAXIES OUT TO THE AIRSTRIP AND TAKES UP POSITION AS THE SQUADRON COMMANDER'S NUMBER TWO.

OK, GANNIC,* LET'S GO.

*THE SQUADRON'S CALL SIGN ON THE R/T (RECEIVER/TRANSMITTER).

THEY ARE IN THE AIR WITHIN TWO MINUTES.

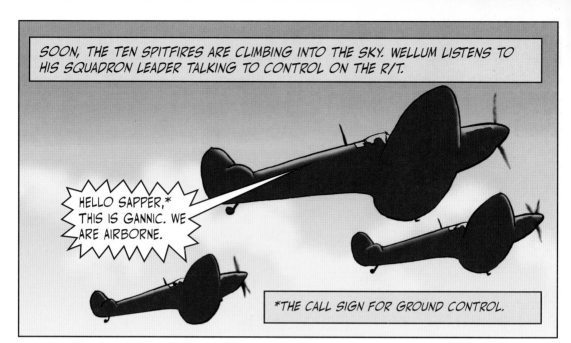

SOON, THE TEN SPITFIRES ARE CLIMBING INTO THE SKY. WELLUM LISTENS TO HIS SQUADRON LEADER TALKING TO CONTROL ON THE R/T.

HELLO SAPPER,* THIS IS GANNIC. WE ARE AIRBORNE.

*THE CALL SIGN FOR GROUND CONTROL.

OPERATIONS CONTROL ROOM.

GANNIC, THIS IS SAPPER. BANDITS* ARE HEADING FOR DUNGENESS AT ANGELS** TWELVE.

*CODE WORD FOR ENEMY AIRCRAFT.
** CODE WORD FOR HEIGHT–12 = 12,000 FT (3,650 METERS).

SAPPER, THIS IS GANNIC LEADER. UNDERSTOOD.

GANNIC LEADER, BANDITS INCLUDE SNAPPERS.* I REPEAT, MANY SNAPPERS.

*THE CODE WORD FOR ENEMY ME-109 FIGHTER PLANES.

AS THEY CLIMB HIGHER, WELLUM TURNS ON THE OXYGEN. THE ALTIMETER READS 9,000 FEET (2,745 METERS). SUDDENLY OVER THE R/T...

SAPPER, THIS IS GANNIC. I SEE THEM. THEY ARE AT ANGELS ONE FIVE, AND THERE ARE HUNDREDS OF THEM!

AS THEY GET CLOSER, THE SQUADRON LEADER GIVES THE ORDER TO ATTACK.

BANDITS ONE O'CLOCK HIGH.* OK, CHAPS, IN WE GO. TALLY-HO! TALLY-HO!

*PILOTS USE A CLOCK FACE TO DESCRIBE THE POSITION OF SOMETHING. TWELVE O'CLOCK IS TO THE FRONT AND SIX O'CLOCK IS BEHIND.

ACHTUNG!
SPITFIRE!

AS THEY CLOSE IN ON THE ENEMY PLANES, WELLUM PICKS OUT A DORNIER TO
ATTACK. BULLETS FROM THE FRONT GUNNER ZIP BY HIS CANOPY.

HE FIRES HIS GUNS, AND BULLETS SMASH INTO THE GLASS NOSE OF THE
GERMAN BOMBER.

AT THE LAST MINUTE WELLUM PULLS INTO A VIOLENT DIVING TURN TO AVOID
COLLIDING WITH THE DORNIER.

WELLUM PULLS UP OUT OF THE DIVE AND ENTERS THE DOGFIGHT AGAIN. HE SEES AN ME-109 AND FIRES A QUICK BURST.

SIX MORE AT ONE O'CLOCK HIGH.

ONE-O-NINES AT FOUR O'CLOCK.

HE MANAGES TO FOLLOW IT, BUT A HURRICANE GETS IN THE WAY AND HE HAS TO BREAK OFF. THE R/T IS NOISY WITH VOICES OF THE BRITISH PILOTS.

BLUE SECTION! BREAK STARBOARD!

HE LATCHES ONTO ANOTHER ME-109, BUT TRACER FROM BEHIND FORCES HIM TO BREAK HARD DOWNWARD, SO THAT THE ME-109 FOLLOWING OVERSHOOTS.

THE DOGFIGHT IS THINNING OUT A BIT WHEN WELLUM SPOTS A HEINKEL III BELOW HIM, HEADING FOR THE COAST.

I SEE YOU.

WELLUM GAINS ON THE HEINKEL AND, IGNORING THE FIRE FROM THE REAR GUNNER, FIRES A BURST INTO THE RIGHT ENGINE.

AS HE BREAKS DOWNWARD TO THE LEFT, THE REAR GUNNER SCORES A HIT ON HIS RIGHT WING.

PING PING

THAT WAS CLOSE!

WELLUM LOOKS OVER HIS SHOULDER BEFORE RETURNING TO THE ATTACK. HE SILENCES THE REAR GUNNER AND SHOOTS UP THE LEFT ENGINE.

THAT GOT HIM.

AS THE HEINKEL STARTS LOSING HEIGHT, WELLUM FIRES ANOTHER BURST FOR GOOD MEASURE. SUDDENLY, HIS GUNS STOP FIRING.

THAT WAS SILLY. I'VE RUN OUT OF AMMO.

WHAT THE...?

I'VE BEEN BOUNCED FROM BEHIND LIKE A COMPLETE BEGINNER!

WELLUM PULLS HIS SPITFIRE INTO A HARD, TIGHT RIGHT TURN...

HE'LL NEVER FORGET HIS FIRST SOLO FLIGHT...

...OR LEARNING TO FLY THE ADVANCED TRAINER, THE HARVARD I, AN UNFORGIVING PLANE THAT KILLED HIS FRIEND NICK.

P5820

NEITHER WILL HE FORGET RECEIVING HIS WINGS...*

AND THEN BEING POSTED TO A FIGHTER SQUADRON BEFORE FINISHING HIS ADVANCED TRAINING.

WELLUM, BE READY TO LEAVE AT O-NINE-HUNDRED HOURS. YOU'RE BEING POSTED.

GOOD OLD MILMAN HAS SEWN THEM ON ALREADY.

*A BADGE WITH TWO WINGS THAT IDENTIFIES AN RAF PILOT.

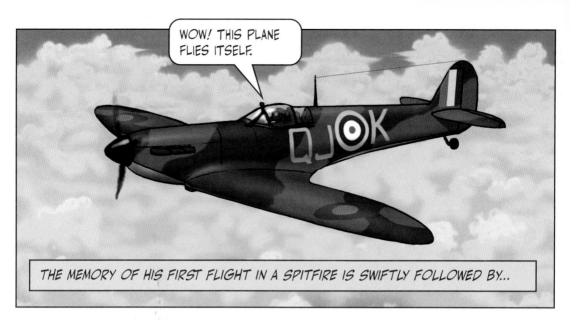

THE MEMORY OF HIS FIRST FLIGHT IN A SPITFIRE IS SWIFTLY FOLLOWED BY...

...MEMORIES OF A NEAR-FATAL CRASH WHEN LANDING IN THE DARK.

WELLUM IS BROUGHT BACK TO THE PRESENT BY THE ME-109 FIRING ITS GUNS AGAIN. THIS TIME THEY MISS HIS PLANE.

IF I KEEP THIS UP HE'LL EVENTUALLY RUN OUT OF FUEL, AND HE'LL HAVE TO HEAD BACK TO FRANCE.

THE TWO PLANES ARE FLYING IN A CIRCLE. THE SPITFIRE HAS THE ADVANTAGE OVER THE ME-109 IN TIGHT TURNS. AS THE TWO PLANES KEEP CIRCLING, THE SPITFIRE GRADUALLY GAINS ON THE MESSERSCHMITT.

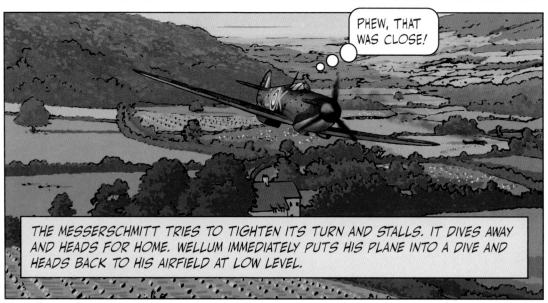

PHEW, THAT WAS CLOSE!

THE MESSERSCHMITT TRIES TO TIGHTEN ITS TURN AND STALLS. IT DIVES AWAY AND HEADS FOR HOME. WELLUM IMMEDIATELY PUTS HIS PLANE INTO A DIVE AND HEADS BACK TO HIS AIRFIELD AT LOW LEVEL.

ARE YOU ALL RIGHT, SIR? I'M AFRAID YOUR PLANE CANNOT BE MENDED HERE. IT'S A TOTAL MESS.

GEOFFREY WELLUM CONTINUED TO FLY FIGHTERS AND SURVIVED THE WAR. HE WAS DECORATED WITH THE DISTINGUISHED FLYING CROSS, AND HIS MEMOIRS OF THE WAR WERE PUBLISHED IN A BOOK CALLED "FIRST LIGHT." **THE END**

LT. RANDALL H. CUNNINGHAM
U.S. NAVY FIGHTER PILOT
1ST FIGHTER ACE OF THE VIETNAM WAR

IT IS MAY 10, 1972. LT. RANDY "DUKE" CUNNINGHAM IS FLYING A PHANTOM F-4J DURING AN AIR STRIKE ON THE HAI DONG RAILYARDS IN VIETNAM. BEHIND HIM SITS HIS RADAR INTERCEPT OFFICER (RIO), LT. WILLIE DRISCOLL. THEY HAVE ALREADY SHOT DOWN TWO ENEMY MIG JETS ON PREVIOUS MISSIONS. SUDDENLY, THERE IS A SHOUT OVER THE RADIO FROM THE PILOT OF ANOTHER PHANTOM...

SHOWTIME ONE HUNDRED,* MIGS ON YOUR SIX,** BREAK RIGHT!

TWO NORTH VIETNAMESE JET FIGHTERS, RUSSIAN-BUILT MIG-17S, ARE DIVING ON THEM.

*THE NAME OR CALL SIGN OF CUNNINGHAM AND DRISCOLL'S PLANE.
**AS IN SIX O'CLOCK (SEE PAGE 21).

CUNNINGHAM TURNS THE PHANTOM SHARPLY TO THE RIGHT, AND THE LEADING MIG SHOOTS PAST.

SWOOOSSSSH

CUNNINGHAM REVERSES HIS TURN AND HEARS THE "LOCK ON" TONE FOR THE SIDEWINDER MISSILE. HE FIRES AND THE SIDEWINDER SHOOTS TOWARD THE MIG.

THE MISSILE EXPLODES, SENDING THE MIG DOWN IN FLAMES.

CUNNINGHAM USES HIS AFTERBURNERS TO PUT SOME DISTANCE BETWEEN THE PHANTOM AND THE OTHER MIG.

THOSE PHANTOMS HAD BETTER GET OUT OF THERE!

CUNNINGHAM HEADS BACK AND FINDS A CIRCLE OF EIGHT MIGS WITH TWO PHANTOMS AMONG THEM.

SHOWTIME ONE-TWELVE, BREAK RIGHT, NOW!

A PHANTOM TURNS SHARPLY IN FRONT OF THEM, NEARLY CAUSING A COLLISION.

CUNNINGHAM IS NOW BEHIND ONE OF THE MIGS. IT MAKES A RUN FOR IT.

ALTHOUGH THE MIG IS TOO CLOSE FOR A MISSILE TO ARM ITSELF, CUNNINGHAM FIRES A SIDEWINDER ANYWAY.

THE SPEED OF THE DIVING MIG IS ITS UNDOING. IT GIVES THE MISSILE TIME TO ARM ITSELF.

FFFFSSSSSHHHHH

BLAM

THE MISSILE EXPLODES IN THE TAILPIPE OF THE MIG.

IT IS CUNNINGHAM AND DRISCOLL'S FOURTH "KILL" OF THE WAR.

SUDDENLY, DRISCOLL SPOTS MIG-21S. THESE ARE SUPERIOR FIGHTERS, AND CUNNINGHAM AND DRISCOLL ARE ON THEIR OWN.

LET'S GET OUT OF HERE!

CUNNINGHAM LIGHTS UP THE AFTERBURNERS AND THEY HEAD BACK TO THE AIRCRAFT CARRIER, USS CONSTELLATION.

HEY, WATCH THIS, WILLIE. I'M GOING TO SCARE THE PANTS OFF HIM.

ON THEIR WAY BACK THEY SEE A MIG-17 HEADING STRAIGHT FOR THEM.

CUNNINGHAM INTENDS TO PASS VERY CLOSE TO THE MIG, A TACTIC THAT MAKES IT DIFFICULT FOR THE ENEMY PLANE TO DOUBLE BACK ON THE PHANTOM'S SIX.

IT IS CUNNINGHAM'S FIRST MISTAKE. UNLIKE THE PHANTOM, THE MIG HAS A GUN ON ITS NOSE...

HE'S FIRING AT US!

CUNNINGHAM PULLS HARD ON THE STICK AND SENDS THE PLANE INTO A STEEP CLIMB.

ZIP

ZIP

ZIP

THEY LOOK BELOW, AS MIG PILOTS TEND NOT TO TAKE ON PHANTOMS IN VERTICAL FIGHTS.

I CAN'T SEE HIM.

SUDDENLY THEY SEE THE MIG CLIMBING LESS THAN 300 FEET (100 METERS) AWAY. THIS IS NO ORDINARY PILOT!

HUH, WHAT'S HE DOING THERE?!

CUNNINGHAM APPLIES THE AFTERBURNERS AND THEY CLIMB FASTER.

IT IS HIS SECOND MISTAKE. HE HAS PUT THEM BACK INTO THE MIG'S GUN SIGHTS.

YIKES, HE'S FIRING AT US!

ZIP

CUNNINGHAM ROLLS THE PHANTOM OVER...

THE MIG DOES THE SAME.

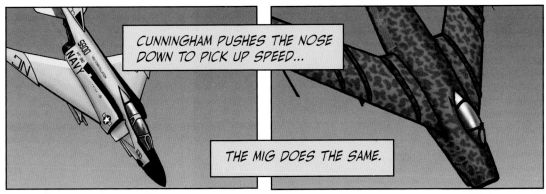

CUNNINGHAM PUSHES THE NOSE DOWN TO PICK UP SPEED...

THE MIG DOES THE SAME.

CUNNINGHAM THEN FLIES AT THE MIG AND ROLLS OVER THE TOP OF IT, WHICH BRINGS HIM BEHIND HIS ENEMY, BUT TOO CLOSE TO FIRE A MISSILE.

THE MIG PILOT HEADS FOR THE PHANTOM, FORCING CUNNINGHAM TO OVERSHOOT. BOTH JETS ARE NOW IN A CLASSIC ROLLING SCISSORS.*

*THE ROLLING SCISSORS OCCURS WHEN EACH JET TRIES TO GET BEHIND THE OTHER BY ROLLING AROUND HIS OPPONENT. THE EXTRA DISTANCE COVERED BY A ROLLING PLANE BRINGS IT BEHIND ITS OPPONENT.

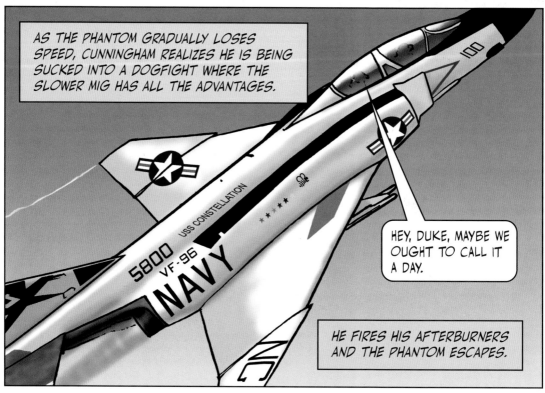

AS THE PHANTOM GRADUALLY LOSES SPEED, CUNNINGHAM REALIZES HE IS BEING SUCKED INTO A DOGFIGHT WHERE THE SLOWER MIG HAS ALL THE ADVANTAGES.

HEY, DUKE, MAYBE WE OUGHT TO CALL IT A DAY.

HE FIRES HIS AFTERBURNERS AND THE PHANTOM ESCAPES.

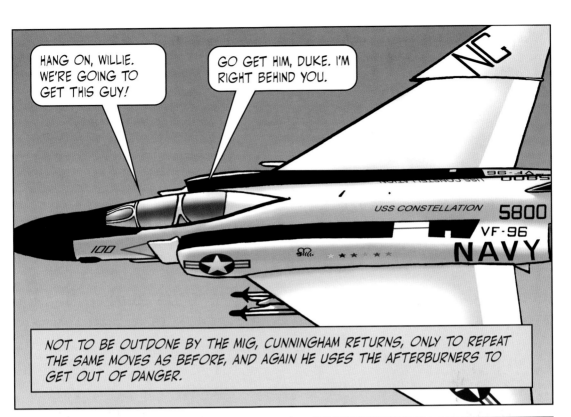

HANG ON, WILLIE. WE'RE GOING TO GET THIS GUY!

GO GET HIM, DUKE. I'M RIGHT BEHIND YOU.

NOT TO BE OUTDONE BY THE MIG, CUNNINGHAM RETURNS, ONLY TO REPEAT THE SAME MOVES AS BEFORE, AND AGAIN HE USES THE AFTERBURNERS TO GET OUT OF DANGER.

AT THE THIRD ATTEMPT, CUNNINGHAM REMEMBERS HIS TOP GUN TRAINING. AS THE PLANES GO INTO A STEEP CLIMB, HE PUTS THE AIR BRAKES ON AND REDUCES HIS THROTTLE. THE MIG SHOOTS OUT IN FRONT OF HIM.

REALIZING THE DANGER HE IS IN, THE MIG DIVES FOR THE GROUND. HE HOPES THAT THE HEAT OF THE GROUND WILL PUT OFF ANY HEAT-SEEKING MISSILE FIRED AT HIM.

CUNNINGHAM PULLS THE PHANTOM OVER AND GETS IT INTO A FIRING POSITION BEHIND THE FLEEING MIG.

I'M GOING TO LET ONE LOOSE ON HIM!

HE FIRES A SIDEWINDER...

SWISHHH

THE MISSILE HOMES IN ON THE HEAT OF THE MIG'S JET EXHAUST.

BLAM

THERE IS AN EXPLOSION.

THE MIG CONTINUES DIVING. SMOKE STARTS POURING FROM ITS EXHAUST. IT HITS THE GROUND AND EXPLODES.

KERBLAM

IT IS CUNNINGHAM'S FIFTH "KILL" OF THE WAR, WHICH MAKES HIM THE FIRST FIGHTER PILOT ACE OF THE VIETNAM WAR.

A SURFACE-TO-AIR MISSILE (SAM) IS FIRED BY THE NORTH VIETNAMESE ARMY.

KWOOSH

THE PHANTOM'S SYSTEMS PICK UP THE INCOMING SAM...

DUKE, WE'VE GOT A SAM ON OUR TAIL!

TO MAKE MATTERS WORSE, THE PLANE KEEPS FLIPPING ON ITS BACK.

EJECT! EJECT!

SUDDENLY, THE TAIL SECTION BRAKES AWAY AND THE PLANE GOES INTO A SPIN.

THEY ARE OVER THE OCEAN. BOTH CUNNINGHAM AND DRISCOLL ARE PICKED UP BY A U.S. NAVY HELICOPTER AND MAKE IT SAFELY BACK TO THE USS CONSTELLATION. LATER, CUNNINGHAM RECEIVES THE NAVY CROSS FOR HIS HEROISM AND AIRMANSHIP ON THIS DAY.   **THE END**

HOW TO BECOME A FIGHTER PILOT

It is possible to become a fighter pilot in the U.S. Air Force, U.S. Navy, or U.S. Marine Corps. You will have competition, though. Out of every 100 people who apply, 50 are selected and about 35 of those 50 get through training. For those who make it, little rivals the fast pace and excitement that this career offers.

REQUIREMENTS

All pilots in these services are commissioned officers, and they must be college graduates. That means you must have a four-year college degree with a high college grade-point average, and you must score well on the service's flight aptitude test.

STEPS TO BECOMING A FIGHTER PILOT

1. Enter an Officer Program. There are several programs, such as the Naval Academy in Annapolis, Maryland, or the Air Force Academy in Colorado; or through college-based ROTC (Reserve Officer Training Corps), or the Air Force OTS (Officer Training School). Admission requires passing a test (very much like a Scholastic Aptitude Test, or SAT), and passing a rigorous physical exam. Excellent eyesight is important.

2. Flight School. Once you have completed Officer Training and received your commission you are sent to Flight School. Here you will learn about meteorology, aeronautics, and principles of flight–flying planes in the air and on simulators. After about a year, if you successfully complete the course, you will be given your "wings" as a pilot. If you finish at or near the top of your class, you will go on to Advanced Training as a Fighter Pilot.

3. Advanced Training. You will learn to fly a jet fighter, and air-to-air combat skills. Survival school is part of the training, too. You will learn how to survive in the wilderness, and how to escape capture if you are shot down. Once advanced training is completed, you will be assigned to a squadron to fly a jet fighter.

Maj. Michael Hernandez,
an F/A-22 Raptor pilot,
discusses preflight operations with
Senior Airman James Douglass, F/A-22
Raptor crew chief. The F/A-22 Raptor is the
newest fighter in the U.S. Air Force inventory.

GLOSSARY

aerobatics Flying maneuvers performed by a pilot.

aeronautics The science of moving through the air.

afterburners Extra burners fitted to the exhaust system of a jet engine to provide increased thrust.

Allied Forces The joint military forces fighting against Germany and Japan during World War II.

avionics Electronics in airplanes.

bounced Unexpectedly attacked by another aircraft.

cockpit The place in a plane where the pilot sits.

commissioned officer A person in the armed forces who has the authority to command a military unit.

composite Made up of more than one material.

dogfight Close aerial combat between military planes.

fatal Resulting in death.

hydraulic system Various systems, such as the wing control surfaces, that are moved by pistons pushing fluid through pipes.

mascot A person, object, or animal that is supposed to bring good luck to an organization.

meteorology The science of the atmosphere, especially the weather.

monoplane A plane with one pair of wings.

novice A person new to a situation or job.

radar (**ra**dio **d**etection **a**nd **r**anging) The system for detecting enemy planes by sending out radio waves, which reflect back from the target to a receiver.

receiver/transmitter Radio that receives and transmits sound.

reconnaissance Military observation of the enemy by scouting.

rudder Vertical control surface at the back of a plane used to steer the plane left or right.

simulator A machine with a similar set of controls to a real machine, that allows people to train without damaging it.

sortie A single mission by an aircraft.

stall A condition when a plane flies too slowly for the pilot to control it, and it stops flying.

stealth The technology applied to some modern military jets to avoid detection by radar.

taxi To move slowly and purposefully along an airfield.

technology Modern scientific knowledge that is applied to engineering.

tracer Bullets that give off a trail of light to mark their flightpath.

variable wing A wing that can sweep back for fast flight.

volunteer A person who freely enlists for military service.

FOR MORE INFORMATION

ORGANIZATIONS

U.S. Air Force
AFROTC Admissions
551 East Maxwell Blvd
Maxwell AFB
Atlanta, GA 36112-5917
(866) 423-7682
Wb site: http://www.afrotc.com/careers

FOR FURTHER READING

Dartford, Mark. *Fighter Planes* (Military Hardware in Action).
Minneapolis, MN: Lerner Publications, 2003.

Hansen, Ole Steen. *The Story of Flight: Military Aircraft of World
War I.* New York, NY: Crabtree Publishing Company, 2003.

Hansen, Ole Steen. *The Story of Flight: Military Aircraft of World
War II.* New York, NY: Crabtree Publishing Company, 2003.

Hansen, Ole Steen. *The Story of Flight: Modern Military Aircraft.*
New York, NY: Crabtree Publishing Company, 2003.

Kennedy, Robert C. *Life as an Air Force Fighter Pilot.* New York, NY:
Children's Press, 2000.

Parks, Peggy J. *Fighter Pilot.* Oxford, England: Heinemann, 2005.

Purcell, Martha Sias. *Pioneer Pilots and Flying Aces of World War I.*
Logan, IA: Perfection Learning, 2004.

Quick, Kathryn. *Blue Diamond.* New York, NY: Avalon, 2000.

INDEX

Web Sites

Due to the changing nature of Internet links, Rosen Publishing has developed an online list of Web sites related to the subject of this book. This site is updated regularly. Please use this link to access the list:

http://www.rosenlinks.com/gc/fipi